AF228464

ALGORITHMS & SEQUENCING

by Teddy Borth

Cody Koala

An Imprint of Pop!
popbooksonline.com

abdobooks.com
Published by Pop!, a division of ABDO, PO Box 398166, Minneapolis, Minnesota 55439. Copyright © 2022 by Abdo Consulting Group, Inc. International copyrights reserved in all countries. No part of this book may be reproduced in any form without written permission from the publisher. Cody Koala™ is a trademark and logo of Pop!.

Printed in the United States of America, North Mankato, Minnesota

052021
092021

THIS BOOK CONTAINS RECYCLED MATERIALS

Cover Photo: Shutterstock Images
Interior Photos: Shutterstock Images, 6 (top), 6 (bottom center), 6 (bottom right), 9 (top), 9 (bottom left), 9 (bottom center), 10, 13, 17, 18 (top), 18 (bottom left), 18 (bottom center), 18 (bottom right); iStockphoto, 6 (bottom left), 14, 21

Editor: Elizabeth Andrews
Series Designer: Laura Graphenteen

Library of Congress Control Number: 2020948286
Publisher's Cataloging-in-Publication Data
Names: Borth, Teddy, author.
Title: Algorithms & sequencing / by Teddy Borth
Description: Minneapolis, Minnesota : Pop!, 2022 | Series: Coding basics | Includes online resources and index.
Identifiers: ISBN 9781532169618 (lib. bdg.) | ISBN 9781098240547 (ebook)
Subjects: LCSH: Algorithms--Juvenile literature. | Sequence controllers, Programmable--Juvenile literature. | Computer algorithms--Juvenile literature. | Mathematics--Juvenile literature. | Algebraic functions--Juvenile literature. | Computer programming--Juvenile literature.
Classification: DDC 005.1--dc23

Hello! My name is

Cody Koala

Pop open this book and you'll find QR codes like this one, loaded with information, so you can learn even more!

Scan this code* and others like it while you read, or visit the website below to make this book pop.

popbooksonline.com/algo-sequence

*Scanning QR codes requires a web-enabled smart device with a QR code reader app and a camera.

Table of Contents

Chapter 1

A Set of Instructions

An algorithm is a set of instructions. **Coders** use them to tell the computer how to do a task.

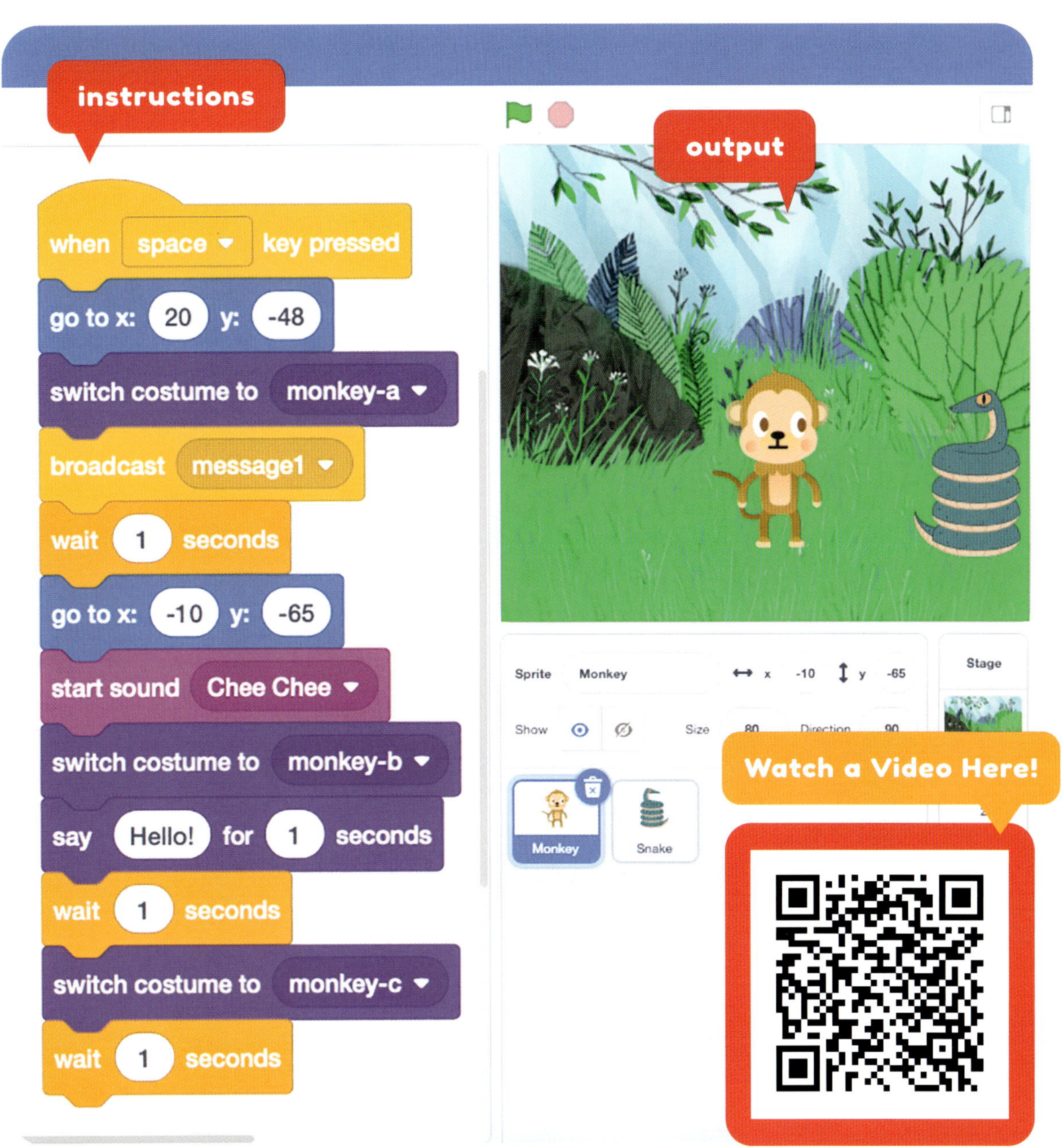

instructions
output
when space key pressed
go to x: 20 y: -48
switch costume to monkey-a
broadcast message1
wait 1 seconds
go to x: -10 y: -65
start sound Chee Chee
switch costume to monkey-b
say Hello! for 1 seconds
wait 1 seconds
switch costume to monkey-c
wait 1 seconds
Sprite Monkey x -10 y -65
Show Size 80 Direction 90
Stage
Monkey
Snake
Watch a Video Here!

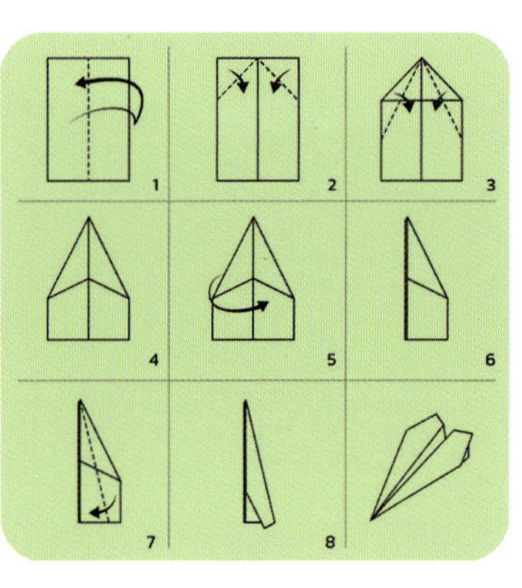

Algorithms are not just for computers. Craft projects use them to make things easier. An algorithm can help you find the best way to fold a paper airplane.

The art of folding paper into shapes is called origami.

Your Very Own Algorithm

Every morning, Aaron has the same routine before school. He wakes up, gets dressed, eats breakfast, and catches the bus. This is his morning algorithm. It gets him to school on time.

Learn more here!

hypotenuse
right triangle

Mia has a certain way she likes to solve math problems. She knows how to find the length of a **hypotenuse**. She goes through it step-by-step. She uses her own algorithm!

Chapter 3

Putting the Steps in Order

The steps of an algorithm need to be in the right order. Dan is going to a new friend's house. His friend gave him directions of when and where to turn. If Dan moves out of order, he will get lost!

Complete an activity here!

underwear
pants

Putting the steps in order is called sequencing. If the steps are done out of order, the algorithm might fail. John gets dressed in the morning. He got the sequence wrong today. Now, his underwear are over his pants!

Chapter 4

Sequencing Your Life

Ollie reads a book. The book has a sequence of events. Things happen in a certain order. If he read the ending first, he would be **confused**!

Learn more here!

Even cereal has a sequence! Katie knows to put the cereal in the bowl first. Then she can add milk. Then she can eat.

Hannah rides the bus home from school. The algorithm is the **route** the bus takes. The bus driver knows the sequence of the stops. Hannah's stop is next!

A good route helps the bus make all its stops in the least amount of time.

Making Connections

Text-to-Self

Have you ever made a sandwich? What is the sequence you used? Does the order of steps matter?

Text-to-Text

Have you read other books about algorithms or sequencing? What did you learn?

Text-to-World

Grocery stores have a lot of items and people. What algorithms do you think are used to stock the shelves? Do you think customers use an algorithm to shop?

Glossary

coder – a person who builds programs or works with computer languages.

confused – to be mixed up or to not understand.

hypotenuse – the longest side of a right triangle, opposite the right angle.

route – a path from start to finish.

Index

Online Resources

popbooksonline.com

Thanks for reading this Cody Koala book!

Scan this code* and others like it in this book, or visit the website below to make this book pop!

popbooksonline.com/algo-sequence

*Scanning QR codes requires a web-enabled smart device with a QR code reader app and a camera.